THE OLD MAN

AND THE

BUTTERFLIES

Story and Photographs

by

Dan Merritt

Acknowledgements

Special thanks to step-granddaughter, Ada Flieschman, sister Elece Pease and friend, Sally Brumfield for reassuring me that this story has appeal for ages 5 to 85. Thanks to dear friend Pat McViegh for her early encouragement and for providing a home in her beautiful yard for a daughter passion vine and its fritillary butterflies. Thanks to daughter Elizabeth for expressing her love of the early photos. Gratitude to Siraj and Sambodhi for inspiration. I am grateful for the editing expertise of Lisa Adams and Steve Kinzie. Thanks to Rebecca Silvers for her help with layout, photo editing, and cover design.

And I am grateful for Nature's daily blessings.

THE OLD MAN

AND THE

BUTTERFLIES

The old man looked closely at the scrawny plant on the wire shelf in front of the grocery store.

The beautiful flower pictured on the side of the plastic pot looked like something from a remote tropical island. The price on the tag had been crossed out three times. It was now on sale for $1.95.

How could such a scrawny little plant get the name "passion vine" and how could it produce such a beautiful flower? he wondered.

He had been searching for something to plant in the raised flowerbed at the little house he had moved into the previous

spring. Most of the small back-yard of the house was covered with concrete. He guessed that the last people who lived in the house didn't want to take care of a lawn or very many flowers.

The flowerbed, extending along the base of the wooden fence, was full of good, rich soil.

In fact, it was so good that the two tomato seedlings he had planted in early spring had grown to the top of the fence and almost to the back door of the house. They had eventually grown over and around the little statue that sat meditating outside his back door.

The two tomato plants had produced delicious little bright yellow pear tomatoes, which he shared with delighted friends and neighbors. But, they were annual plants, so they only lasted one year.

He was happy to have found the passion vine because it was a perennial plant—one that could be healthy and beautiful year after year, maybe even for the rest of his life.

At home, the old man put away the groceries, placed his little plant on a chair next to his desk, and sat down at the computer to learn how to grow passion vine plants.

In his little storage shed, he found a small blue ceramic pot. He mixed soil with some compost from his worm bin and filled the

pot with the rich mixture. He made a small depression in the middle and carefully transferred the plant to the little blue pot. He watered it well and placed it outside in a shady spot so it could adjust to its new home.

During the next two weeks, the old man checked the little plant every day and carefully watered it when the soil felt dry.

After three weeks, he was happy to see tiny new leaves and long, coiled, thread-like tendrils reaching out from the ends of the branches. This told him the roots were healthy and beginning to bring nutrients into the plant from the soil, so he transplanted it into the flowerbed.

He continued to water and watch the little plant as it grew into a large vine and spread along the wooden fence.

He was very pleased, one day, to see several light green flower buds on the plant.

The next time he looked, one of the buds had opened, revealing a beautiful flower. Ten white petals, glistening with droplets of dew, surrounded its delicate halo of lavender. As he admired the new flower, two honeybees flew in and landed on it.

He watched them feeding on the nectar and collecting pollen in the tiny pollen baskets on their back legs. He felt happy, knowing the bees and the passion flower had found each other and that the bees would be taking nectar and pollen back to their hives to feed other hungry bees.

A few days later, while sitting at his desk beneath the east-facing window, paying bills or some such thing, the old man was surprised to see a small shadow flicker across the sunlit desktop in front of him.

He looked up, out the window, but saw nothing moving.

After a few minutes, the flickering shadow came by again. This time, he looked up quickly and saw two small flashes of orange color dancing in the air.

They were butterflies!

The old man dropped his pen and stepped quickly through the

door into the backyard. He watched the two butterflies flit back and forth over the fence until they came to rest on the passion vine, right in front of him. The two butterflies stayed close together, facing each other, for a long time before flying away.

Early, the next morning, after eating a quick breakfast, the old man stepped out the door, his cup of cocoa steaming in the cool morning air.

Would the two butterflies come back? he wondered.

They *had* come back.

This time they were resting quietly on a small vine with their tails together. A bee sat nearby, too busy drinking nectar from a grateful flower to notice.

"They must be mating," the old man said to no one in particular.

He sat for a long time in his patio chair, sipping cocoa, smiling, and watching the butterflies, as more bees came and went.

I wonder what kind of butterflies they are. They resemble monarch butterflies, but they are a bit smaller and have patches of bright silver on the underside of their wings.

A little Internet research helped him identify them as

gulf fritillary butterflies—also called passion butterflies.

Watching the "butterfly and bee show" became a regular

morning ritual for the old man.

One day, he noticed one of the butterflies sitting on a leaf, with

the little tail end of its body pressed down on the surface. After

a few minutes, it flew to another leaf and did the same thing.

The old man looked closely at the leaves and noticed that the

butterfly had left a tiny whitish-yellow bump on each leaf it

had visited.

Adjusting his glasses, he

peered down and squinted

his eyes at the tiny fleck.

"It must be an egg!" He ex-

claimed. From his research

he remembered that, while

adult fritillary butterflies feed on nectar from many different flowers, the caterpillars can only eat passion vine leaves.

He even found some eggs on the coiled tendrils of the vine.

He tried different types of lighting to see the eggs better.

He brought out the magnifying glass from his desk drawer. He loved looking at tiny things through the big lens.

To his surprise, he saw that each tiny egg resembled a miniature corncob!

He wished he could see them even closer, in more detail.

"Wait!" the old man exclaimed. He'd just remembered the used microscope he'd bought last year on eBay. He had gotten it to look at the tiny creatures swimming in pond water that he'd collected from the neighborhood park.

He took the big grey microscope from his closet and set it up at the kitchen table, along with some glass slides, small plastic dishes, and forceps.

Now, he was ready to peek into the hidden world of the tiny eggs.

The old man spent several days examining the eggs left by the female butterfly on the leaves and tendrils of the passion flower vine.

He returned faithfully each day to examine the eggs for any changes.

If they are butterfly eggs, he thought, *there will be caterpillars hatching from them before long.* Questions raced through his mind. *How will the baby caterpillars get through the outer wall of their eggs? Do the eggs crack open like a chicken egg? What will the caterpillars look like--will they be bright orange and black like their parents?*

The answer to his first question came the next day, when he spotted an empty butterfly egg case with a hole in its side. It hadn't cracked open like a chicken egg. Instead it looked like something had eaten its way out!

His second question was answered the next evening when he

was peering through the microscope and was greeted with the sight of a tiny caterpillar crawling out of a hole in its egg case, across the surface of a leaf and down over the leaf's edge, leaving its empty egg case behind.

The tiny caterpillar was orange with a black head and black spikes emerging from each segment of its body.

A few days later, the old man spotted an-

other tiny caterpillar crawling across a nearby flower bud.

He soon spotted other caterpillars crawling along the vine tendrils.

As they grew, their orange and black stripes became brighter.

He also saw that each time they outgrew their skin, they simply crawled out and left the old skin behind—a slender, spiny, black and orange tube. Their growing bodies, which had been waiting beneath the old skin, could now expand within fresh, colorful new skin.

Watching the caterpillars crawling about, eating the passion vine leaves, the old man wondered if he would get to see them grow up and turn into butterfly adults—with their broad gold and silver and black wings, long legs, and two elegant antennae.

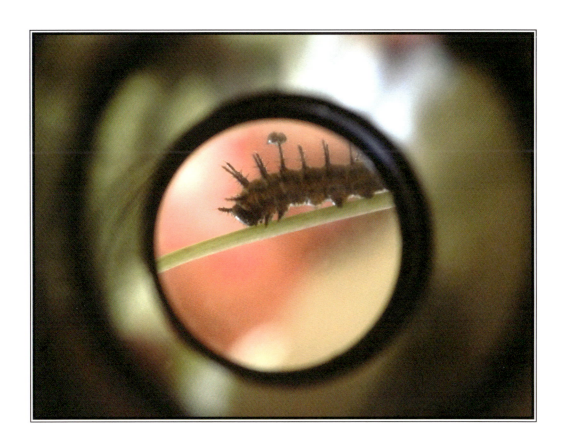

23

23

Into the days of midsummer, he watched the caterpillars make their way along branches, chewing the leaves of the passion vine. He was glad the vine was large enough to provide plenty of food for the hungry caterpillars.

He knew that butterflies went through four stages of *metamorphosis* in their lifetimes—egg, caterpillar, chrysalis, and adult—but it still seemed unfathomable that these spikey-skinned, wiggling little creatures could become passion butterflies.

Two of his favorite caterpillars were best buddies. They were together *all* the time.

One day, the old man discovered one of these caterpillars hanging upside down from a vine stem. Its friend appeared to be watching from the top of the vine stem.

When he looked closer, he could see that the caterpillar had prepared a tuft of fine silky threads to anchor itself by its tail to the underside of the vine.

Soon, the caterpillars seemed to be every-where.

Some were still moving about.

Others were just hanging out.

Many were suspended by their tail ends from vines or fence slats.

They seemed to be the picture of patience.

The old man watched one caterpillar crawl across the top railing of the fence, while others hung directly below—

inside the fence—like ornaments, dangling in tiny windows.

He continued watching the little J-shaped caterpillars, each waiting patiently to shed its skin one last time before transforming into a chrysalis (also called a *pupa*) and to begin the next stage of its life.

During the early stages of this metamorphosis, he watched the

back part of the body keep its bright, striped patterns for a time,

while the front part of the body turned gray and then tan, as it

began to bulge out on both

sides.

When the old man looked

closely, he could see the out-

line of the wings beneath the

surface!

Gradually, the entire surface of the chrysalis came to resemble

a dry leaf.

After hanging quietly for about a week, the chrysalis began to move! The creature within flexed its little body, curving up toward one side, coming parallel to the vine stem then curving down and up toward the other side like a tiny pendulum. Looking more closely at the surface of the chrysalis, he noticed two rows of small bumps extending as parallel lines beginning near

the tail end and diverging near what he assumed would become the head.

He wondered what adult structures might lie beneath these two identical ridges.

Meanwhile, the old man waited eagerly for the first sign of an orange and black butterfly bursting through the outer wall of the chrysalis.

And then one morning it happened. "Hooray!" he shouted—

loudly enough for the next-door neighbors to hear. "There you are!" Welcome!" The old man saw the dry skin of a chrysalis being split apart by the bulging, striped wings of the determined adult butterfly inside.

He set a timer on his cell phone to remind him to check this chrysalis and any others every half-hour. He was determined not to miss seeing the butterfly's amazing birth into this bright new world.

Nearby he found another butterfly emerging from its chrysalis casing. It had successfully freed its abdomen and tail end and the wings

were out but not yet unfolded.

He noticed that the antennae were still inside the chrysalis case. In fact, when he looked closely he could see that the butterfly was holding onto the chrysalis case with its back legs and pushing against it with its front legs, as it strained to pull its antennae out of the two tiny tunnels that lay beneath the pair of bumpy ridges he had noticed on the surface of earlier stage pupas.

When he looked more closely he also noticed, between the two black antenna stalks, two small tan-colored strands extending from the middle of the butterfly's face into the chrysalis case.

These were the two halves of the long proboscis that zip together to make the sucking tube the butterfly uses for sipping nectar and water.

After a few minutes of struggle, the butterfly swung its head over to the left side of the chrysalis case — giving it more leverage against the right side. With another push the right antenna pulled free of the chrysalis case and the two halves of the proboscis followed.

The butterfly curled and straightened the two proboscis halves again and again until they fused together into one tube and curled up neatly beneath its large orange eyes.

The butterfly rested for a while

as its crumpled wings unfolded and expanded in the midday sun. After about half an hour it crawled to an open place on the vine, stretched its wings up and down a few times then fluttered across the yards and over the houses of the neighborhood.

As the old man watched other caterpillars go through metamorphosis, he tried to visualize what was happening inside their little bodies.

He had learned that while the caterpillar is hanging quietly upside down by its tail, the solid flesh inside gradually liquefies; its internal organs and tissues melt into a kind of liquid slush.

This amazing liquid slush nourishes and awakens special little clusters of cells inside the caterpillar called *Imago cells*. (*Imago* refers to the adult stage of butterflies and other insects.)

These groups of imago cells in the caterpillar act like tiny seeds, waiting to sprout and grow into the butterfly's four beautiful wings; its six legs –includ- ing the four long graceful back legs and the two short forelegs that fold up against the "chest"— its two big eyes; its coiled nectar-sipping mouth tube; and its two long graceful antennae topped with the tiny white-striped knobs.

The old man was astonished: inside every baby caterpillar— even before it chews its way out through its egg case—lie the "seeds" of its future adult body parts! These cell clusters wait patiently as the caterpillar crawls about the passion vine, chewing leaves, growing and molting again and again until it attaches by its tail, sheds its spiny skin one last time and gradually transforms into the pupa then into the adult butterfly inside the parchment-like chrysalis case.

After emerging from the chrysalis case, the new butterfly holds on with its strong legs while its wings unfold and fill with fluid.

Meanwhile, any leftover liquid slush passes out of the body through the anus, sometimes forming a small reddish-orange pool on the ground or among the leaves below.

The old man was outside puttering with his potted plants one afternoon when he noticed something which reminded him things don't always turn out the way we might expect or wish they would.

He had discovered an unusual looking butterfly sitting inside the edge of one of his flower-pots. Its wings, abdomen and legs appeared to be normal but its little face was hidden behind a parchment mask!

He could see that its eyes, proboscis, and antennae were still partially embedded in a portion of the butterfly's old chrysalis case.

How could this happen?

For an answer he searched among the passion vines along the

fence until he found an empty chrysalis case left by a different passion butterfly.

There, still attached to the head end of the dry chrysalis case, he saw the long triangular piece that the unfortunate butterfly in his flowerpot had been unable to leave behind when it had emerged from its own chrysalis case.

The old man returned to the masked butterfly for a closer look. He placed his hand in front of the little butterfly and it crawled onto his finger.

When he lifted it and looked more closely, he saw that the eyes were completely covered by the chrysalis casing.

The little white-striped bulbs at the ends of the antennae were exposed but the antennae had been unable to reach out and extend to their full length above the butterfly's head. The two halves of the proboscis had not fused together to form the feeding tube.

A sadness arose within the old man. He wanted to help the little butterfly by trying to peel away the mask, but he could see it was too tightly wrapped around the delicate antennae, eyes and proboscis for his large rough hands to remove without great damage and probably pain to the little butterfly.

With a sigh, the old man returned the masked butterfly to its resting place in the flowerpot and went about quietly caring for his other plants.

Autumn was drawing to a close as the old man sat on his patio, absorbing the warming morning sun and watching a pair of adult butterflies, who seemed to be dancing—a flickering, spiraling dance—in the air above the passion vine. He knew they would soon be mating and the female would lay new eggs on leaves and tendrils of the passion vine, which now stretched several yards wide, almost blanketing the wooden fence.

He recalled, with fondness, the scrawny little vine he had planted the year before. He smiled, with gratitude, for the opportunity he had been given to watch the passion vine grow and make its beautiful flowers, and for the gift of the passion butterflies.

How wonderful it had been to be present and able to watch the butterflies going through each stage of their lives.

As he sat musing, he noticed a small patch of familiar orange and black on the rough, grey concrete of the patio.

He turned in his chair and, leaning down, asked the butterfly, "Are you OK?"

Bending down for a closer look, the old man could see it was moving, but very, very slowly and seemingly with great difficulty.

He saw that its wings were drooping, the outer margins of the wings dusting the concrete as it moved along. It seemed to be crawling more than walking, with its little legs barely able to carry the weight.

Its outstretched wings quivered each time they lifted off the concrete.

The old man was surprised when the quaking little butterfly turned and began crawling directly toward him.

Without thinking about it, he put his hand, palm down, in front of the butterfly.

The butterfly didn't hesitate. It crawled slowly, but delib- erately, up, onto the back of the old man's hand.

It crawled to the old man's wrist, rested for a bit, then

continued crawling up his arm, onto his shirt sleeve.

At this point, the old man stood, stepped over to the passion

vine, and lifted his arm up into the leaves. He held it there, so his little passenger could easily climb off his arm and into the protective foliage.

But the butterfly ignored the tendrils and leaves lying across the shirtsleeve and continued to crawl up the old man's arm, toward his shoulder.

"OK, I guess you know what you want," chuckled the old man.

He stepped back from the vine, sat down, and said, "I'll sit here

as still as I can until you get to where you want to be."

The butterfly continued crawling up and up until it reached the old man's shoulder.

After arriving at the shoulder, the butterfly turned and faced out, across the patio.

"Do you want to fly, but you are too weak?" asked the old man.

The butterfly stayed. It seemed to the old man that it was resting comfortably, unafraid and in no hurry to be going anywhere else.

As they sat together, feeling the breeze and hearing the chimes ringing from beneath the awnings of the back porch, the old man felt a bittersweet blend of sadness and appreciation— sadness that the old butterfly seemed to be nearing the end of its life and appreciation for the closeness he felt to this beautiful little being.

For a moment, he wished they could speak to each other. Then he realized their mutual understanding and appreciation was beyond words. A stillness had settled between them. They were sharing a beautiful, consensual silence. Nothing more was needed, or possible.

After what seemed just the right amount of time, the old man

quietly rose to his feet and stepped over to the passion vine.

He very carefully lifted the butterfly from his shoulder.

He parted the vines and placed the butterfly gently inside,

among the leaves and tendrils.

"Namaste," the old man whispered, through pressed palms.

"Namaste."

flickering shadows
dancing across a desktop —
tender rendezvous

CPSIA information can be obtained at www.ICGtesting.com
Printed in the USA
BVIW12n1926110718
521400BV00004B/6